Z. H Mason

A General Description of Orange County, Florida

Its Soil, Climate, Health, Productions, Resources, and Facilities...

Z. H Mason

A General Description of Orange County, Florida
Its Soil, Climate, Health, Productions, Resources, and Facilities...

ISBN/EAN: 9783744757102

Printed in Europe, USA, Canada, Australia, Japan

Cover: Foto ©Suzi / pixelio.de

More available books at **www.hansebooks.com**

A DESCRIPTION

OF

Orange County,

FLORIDA.

SOUTH FLORIDA

Real Estate Exchange.

OFFICE IN CITY HOTEL BUILDING,

SANFORD, ORANGE CO., FLA,

offer for sale

ORANGE GROVES,

TOWN LOTS,

Farms AND Market Gardens

and unimproved

PINE AND HAMMOCK LANDS

in lots from

1 TO 1,000 ACRES.

J. J. HARRIS & CO.

J. J. HARRIS, Speaker of the State Assembly.

S. B. HARRINGTON, for six years Deputy Assessor.

GEO. E. SAWYER, County Commissioner and Director of the State Fruit Growers' Association.

A General Description

—OF—

Orange County, Florida.

—ITS—

Soil, Climate, Health, Productions, Resources,

—AND—

Facilities of Transportation.

By Z. H. MASON, M. D., of Apopka.

PUBLISHED BY THE

Orange County Immigration Society.

Dr. Z. H. Mason, President, Rev. W. G. F. Wallace, Vice Pres't.

R. G. Robinson, Secretary.

APOPKA, FLA.

ORLANDO, FLA.:
PRINTED BY MAHLON GORE,
ORANGE COUNTY REPORTER.

INTRODUCTION.

The committee appointed by the Orange County Immigration Society to prepare for publication a pamphlet descriptive of Orange County, respectfully submit the following pages from the pen of Dr. Z. H. Mason, believing that they will prove trustworthy and reliable as a source of information concerning the Soil, Climate, Products, Resources and Advantages of the County.

The committee beg leave to take this opportunity of expressing the obligation they are under to Dr. Mason, who, at their request, assumed solely the labor and responsibility of authorship.

Dr. Mason has been well known for many years as an able and conservative writer on Florida, and brings to his aid more than twenty years of residence and experience in semi-tropical fruit culture and farming, and is thus able to speak advisedly of the resources, and of what can be profitably grown in this section of the State. To use his own words, he has in the following pages "confined himself" to a plain statement of *facts* without embellishment, believing that inquirers want *truth* not *fiction*."

The article "Orange County as a Home for Invalids," is worthy the careful consideration of those whose health require that they should seek a climate with mild winters and pleasant summers.

Information not contained in this pamphlet, can be had by addressing the Secretary or either of the officers of the Society.

All letters *containing return postage* will be promptly answered. It is requested that inquirers write their questions on sheets separate from their letters and *leave space under each question for the answer.*

W. G. F. WALLACE,
R. G. ROBINSON,
Committee of Publication.

ORANGE COUNTY.

Topographical Description and Soils.

Orange County is in that portion of the State designated as South Florida, and is situated on the Peninsula, which averages in width about ninety miles, is midway between the Gulf of Mexico and the Atlantic ocean, and extends from Lake George on the north to Lake Winder on the south, occupying more than a degree of latitude. On the east the boundary is the St. Johns river for a distance of one hundred miles. The western line passes through the great Lake Ahapopka, and the famed Oclawaha river is a portion of the line between Sumter County, and on the south are Brevard and Polk Counties.

Many of the principal streams on the Peninsula have their source in the County and empty into the Atlantic, Gulf and Lake Okeechobee. The elevation of the lands makes this County the water-shed of this portion of the State. The general contour of the country is rolling, this being more especially the case in the interior.

On the St. Johns are large bodies of land termed Savannas, are alluvial, and subject to overflow during high waters. They furnish fine grazing for cattle. There are also many Prairies of various sizes which are only useful for pasturage. Swamp lands are on the borders of streams. The country is dotted with lakes from an acre in extent to those covering thousands of acres. Among the latter is Ahapopka, said to be the second in size in the State. These lakes are stocked with a variety of fish, and that great delicacy, the soft shell turtle. On the borders of many of the lakes and prairies are bodies of rich hammock lands, timbered with live oak, water oak, cherry, hickory, magnolia, red bay, (called Florida mahogany) cabbage-palm and a variety of vines and shrubs. In complete contrast are the scrub lands, covered with a sparse vegetation consisting mainly of scrub oaks and pines, tietic and gall berries, with a soil utterly worthless. The flat-woods are timbered with pines, have an under strata of hard pan, and during the rainy season are often covered with water. There is but little of this kind of land fit for cultivation or fruit raising.

The larger proportion of the lands in the county are high and rolling with a natural drainage and are timbered with pine, several varieties of oak, persimmon, and often wild cherry. In many parts of the county these lands have a clay subsoil. A large proportion of our orange groves are upon these lands, and the successful result of their culture shows the adaptation of such soils to the production of citrus fruits. Pine lands will produce fair crops of corn, cotton, oats and sweet potatoes when given what would be considered at the north a very light dressing of manure. But few have tested their productive capacity by the aid of large quantities of vegetable or animal manures. Where the experiment has been made the result has proved that it will pay. It has been truthfully said in regard to our soils, "That were one-half or one-quarter the expense in fertilizers put upon them that is yearly put in the lands north and in New England, they would excel anything known in the agricultural line." The stately pines commingled with a variety of oaks and undergrowth (where the fires are kept off) attest their fertility. They differ from the pine lands of Michigan and other sections, which are entirely worthless when the timber has been taken off. Our semi-tropical climate has much to do in developing the latent fertility of this kind of soil when brought into cultivation. Most of our hammock lands are naturally fertile, producing large crops of corn, sugar cane and cotton for a long series of years without manures. From the foregoing brief description it will be seen that in Orange County can be found almost every conceivable variety of soil, adapted to the growth of nearly every crop that may be desired.

Timber.

The larger portion of the lands are timbered with yellow and pitch pine, with black jack, turkey oak, post oak and persimmon. The hammock land growth is hard wood, consisting of live oak, water oak, magnolia, cherry, white bay, red bay hickory, gum and cabbage palm. In the swamps are found cypress of large size, cedar, wahoo or linden, maple, ash and oaks. The cypress is especially valuable for shingles, barrels, tubs, buckets, etc.; and as railroads are extended into the interior our hickory, cherry and red bay will find a good market. Even now pine timber is being carried from the interior by railroad to Sanford.

Field Crops.

Corn is one of our most important grain products, both for bread and horse feed. When planted upon hammock land the yield is from twenty-five to fifty bushels per acre; upon pine land the average is about ten bushels. This may be considered a small crop by a western farmer, but it must be remembered that the friable character of our soil enables us to cultivate three acres with less labor than one in those sections where the soil is stiff. This crop is ready to be housed by the 1st of August, giving the Florida farmer a long season in which to attend to other crops. Upland rice is successfully cultivated; an acre patch yielding enough to amply supply a large family with this almost indispensable article of food. Sweet potatoes are largely planted by every farmer. Two plantings are made, one in May and June and the second in August (termed standovers), they being ready to dig the following March and April. Sugar cane is a crop that pays well for the labor expended. It can only be raised upon land naturally rich or that which has been heavily manured, as it is an exhausting crop. The first year's crop is called plant cane, which continues to send up shoots for four or five successive years, and is termed rattoon cane, so that upon rich soil one planting is sufficient for that length of time. The average yield per acre is from 350 to 400 gallons of syrup or 2,000 pounds of sugar, for which there is generally a home demand at good prices. Sea island cotton is planted more or less by every farmer, as it will purchase his supplies or bring cash. The usual yield on pine land is 400 pounds seed cotton, and upon hammock about 800 pounds. Red rust-proof oats are sowed in the fall and generally turn out well. Other kind of oats are liable to be destroyed by rust. This also prevents the raising of wheat. Irish potatoes are planted in December and January for northern markets. Watermelons are raised in large quantities for shipment. They have a reputation for size and luciousness. The first crop ripens about the middle of May, some years earlier. Tobacco, where it has been tried, has proved a success. By using fresh seed every two years a good article of Cuba has been produced. A plant crop, and often two sucker crops, are obtained. Cassava and arrow root yield heavily. There is a factory at Wilcox, where a fine article of farina is prepared from these roots. The cultivation of these starch producing plants should be entered into largely by our people. We have the assurance that where they are raised in sufficient quantities, factories will be established at convenient points.

Vegetables.

The growing of vegetables for market is being largely entered into by many of our citizens, and the result is so encouraging that many others express their intention to plant largely this fall and winter. The principal vegetables cultivated for early shipment are snap beans, peas, cucumbers, tomatoes, Irish potatoes, cauliflower and cabbage ; of the latter, as much as $600 has been received for the product of an acre. Bermuda onions are a good crop, do well here, and sell north in April and May at $9 per barrel. All the usual vegetables raised at the north are grown here. One party exhibited at the County Fair last February thirty-six varieties of his own raising near Apopka. A correspondent of the "Savannah News" writes in regard to the exhibit at the State Fair, held at Jacksonville last January : "As fine vegetables as ever were displayed at a fair can now be seen on exhibition at the fair grounds. Oranges as fine as ever were seen, are numerous, and many of them plucked from the trees during the past week. The cold did not seem to effect them in the least. The majority of these " unfrosted" oranges, as they are called, come from Orange, Volusia and Putnam Counties. The largest turnips (rutabaga) that have ever been exhibited at a State Fair are to be seen here to-day. Orange County is ahead of all other counties in the vegetable line, and is the best advertisement that could be had of the climate and soil. The cauliflower is equal to any seen in Fulton Market, New York ; one barrel of it brought in Philadelphia last week ten dollars. An Orange County gentleman informed me to-day that he sold seven hundred and fifty dollars worth of this food vegetable last year. Magnificent specimens of ripe tomatoes, lettuce (very large heads), potatoes, turnips, &c., from Orange County are exhibited." We can put vegetables into market three weeks to a month earlier than States north of us, ensuring high prices for early products, and for that length of time a control of the markets.

There is one requirement necessary to make gardening a success, and this is common to all countries—a rich soil, if not naturally such, must be made so. Orange County affords in the operations of the garden a never ending round of profitable work: in each month something is required to be done—something appropriate to the season. There is not a single week in the whole year in which something palatable and wholesome may not be gathered for the table, and if the family does not possess these comforts the fault must be in the negligence or want of skill in the cultivator, and not in the climate or soil.

Fruits.

The cultivation of citrus fruits is rapidly extending over the county; even now a large portion of our citizens are engaged in this business. Every year the area in acres increases, and this development will continue until all available lands are covered with fruit farms. Past experience demonstrates that a considerable portion of our lands are adapted to the orange, lemon, lime, tangerine, and other varieties of this class of fruits. The climate being so mild and uniform gives the fruit full time to thoroughly ripen, and makes this industry one that can be depended on. All kinds of land are not suitable for orange growing; the following should be avoided, as swamp, scrub, and flat pine lands with a hard pan subsoil. The fruit succeeds upon both hammock and pine land. A large proportion of the groves are necessarily planted upon the latter, as it largely preponderates in the county. The advantage of hammock is that the land is naturally rich, though it is expensive to clear, while the pine lands will require manures but can be cheaply prepared for planting. Lands set out in oranges pay a heavier per centum on the investment than anything that can be put upon them. Five year old trees budded come into bearing in from one to two years, and rapidly increase in fruit production. Sixteen year old trees, properly cultivated and manured, should averge 800 oranges; many trees exceed this, and some old isolated trees have produced 10,000; such cases are exceptional, and cannot be used as a basis for computing the profits of a grove.

The apple is not suited in this latitude. The tree becomes an evergreen, blooms nearly all the year, becomes dwarfed, and the fruit is small and worthless. A few trees of an early variety are said to have matured fair fruit. Pineapples do well in some portions of the county; they require a slight protection in the winter. They come into bearing in about eighteen months from the shoot or slip. The Le Conte pear is being tried, but it is too soon to determine whether it will prove a success in this climate. The Japan plum and Japan persimmon are both raised here, and are of fine flavor. The first is a decided success; the second will require further time to decide upon its adaptability. The plum grows wild, and some of the varieties are scarcely inferior to many of the cultivated sorts. There is no doubt but that the better varieties can be budded upon the wild stock. Peaches do well, especially where there is a clay subsoil near the surface. The crops, however, are light; the mild winters cause them to bloom through the season. Varieties of China peaches are being tried with satisfactory results. Figs of various kinds grow luxuriantly with but little care, and are readily propagated from cuttings, which bear the second year. The ripe fruit is rich and luscious, and should be found in its season upon every table, being wholesome and nutritious. Every house has its banana patch, furnishing that well known fruit, and is propagated from the shoots which spring up around the mother plant. The broad, long leaves, showing a semi-tropical growth, impart an air of coolness and comfort to the house, as well as being highly ornamental. Guavas are tender, yet in this section seldom fail to yield a supply of fruit that not only furnishes the famed guava jelly, but many a rich dish for the table.

Berries.

Among the small fruits the strawberry stands pre-eminent. It is easily cultivated and bears abundant crops. It requires a moist and fertile soil, and begins to ripen fruit in December, and if freely watered continues in bearing for six months. Wilson's Albany is considered the best and surest bearer in this section. It pays well to send the early crop to a northern market. Blackberries grow on the banks of streams, bay-heads, and in low hammocks. The running variety, or dewberry, grows in old fields. Raspberries, currants and gooseberries have thus far proved a failure: the two latter mildew and rot. Blueberries, huckleberries and bearberries grow on the flat lands. A large variety of wild grapes grow luxuriantly in the woods. But little attention has been given to any kind but the Scuppernong, which is cultivated more extensively than any other variety. One vine on good soil will cover an arbor extending over an acre. The vine does not require pruning, is free from the attack of insects, and the fruit makes a good wine.

Forage Plants.

There are many kinds of wild grass growing in the hammocks, prairies and flat woods that sustain herds of cattle, but all the efforts thus far made to grow the cultivated grasses of the North and West have failed; they seem to do well planted in the fall, but are killed out by the hot sun in May. A few years ago C. Codrington, Esq., editor of the Florida Agriculturist, introduced from the Island of Jamaica the Guinea grass so highly prized there. It furnishes a large amount of forage, grows in large bunches, can be cut for hay or for soiling; it is a great acquisition. Para grass is found to be well suited for planting upon low, moist land, makes a good pasturage, and is highly relished by all kinds of stock. All varieties of millet do well, especially the cat-tail, which can be cut for feeding green every ten days. Some parties are giving the Bermuda grass a trial as pasturage; it does not grow high enough to cut for hay. Rice straw is highly prized as forage, being far preferable to that of oats or rye. The cow pea vines also make a hay that is highly prized, horses and mules being fond of it. Where the land is free from sand and grass spurs, crab-grass, which grows luxuriantly upon cultivated lands, makes a hay equal to timothy. Rust proof oats, planted in October and November, almost invariably do well.

Manures.

Our cattle are not kept up, but are turned out on the natural range, where they live both winter and summer, and will not eat dry feed unless taught to do so when they are calves; therefore all the animal manure we save is from horses and mules. The only way in which we utilize our cattle for manuring is in the spring when they are gathered for the purpose of marking the calves; we pen them at night upon such land as we wish to enrich. A few weeks of such treading fits the land for the production of vegetables or sugar cane. Nature has provided us with an unfailing source from which to draw our supplies of a material which experience has proved to be the very best that we can apply to our lands when properly composted. This is found in *muck*, which is abundant in swamps, bay-heads, borders of lakes and praries. Experiments made with fertilizers upon pine land indicate that swamp muck is one of the best manures for sugar cane. When composted with lime or potash it becomes an efficient manure for almost any crop. Some compost their stable manure with muck, pine straw, cones or wire grass and weeds. Lime stone, sufficiently pure for agricultural purposes, is found at Rock spring, one of the heads of the Wekiva river, and there is a deposit on the Wekiva, one mile from Clay spring. These can be burned into a fair lime. Oyster shell lime is laid down at Sanford at a cost of ten dollars per ton in barrels. Special manures can be purchased at the agencies in this county at manufacturers' prices with freight added. An efficient and cheap method of manuring is to plant the land with cow peas (which have been well called southern clover) at the rate of one and a half bushels per acre, and turn under the vines.

Building Material.

The mild winters of this section, where the thermometer seldom gets down to thirty-two degrees, precludes the necessity of building tight, expensive houses, such as are necessary, not only for comfort, but for safety in cold climates. Our residences are constructed of pine lumber, and of such forms as are calculated to make them comfortable during the long summers. A heavy item of expense is saved in this country; we do not dig cellars, as they are not needed. Lumber is sold at the numerous mills at from $8 to $20 per 1,000 feet, according to quality. For planed lumber the price is greater. Clay for making brick is found in many places in the county, and prices are reasonable. There is in the center of the county a sand stone that will stand fire and makes excellent hearths and chimney jambs. At present all the saw mills are behind with their orders at least six weeks. The rapid immigration now going on requires additional mills to supply the increasing demand for lumber.

Water.

Good free stone water is obtained from wells which vary in depth from eighteen to fifty feet. The water is not what would be called cold at the north, but a person soon becomes accustomed to it. There are numerous springs, especially about bay-heads and on the borders of hammocks; some are free stone, others sulphur, while others again are chalybeate, and these different springs are often within a few feet of each other. Clay spring, one of the heads of the Wekiva, affords a sufficient quantity of water for steamboats to come into the spring. The water is white sulphur, and has proved beneficial in rheumatism and skin diseases. The same kind of water is found at Rock spring, which bursts out of a cleft in the rocks at the base of a cliff some twenty feet high.

Lands.

A large proportion of the best lands belonging to the government subject to homesteading or entry have been taken; there are, however, many homestead claims where parties have not complied with the law, that can be contested. The state lands, except an occasional isolated fraction, have been purchased. Therefore most of the desirable lands are in second hands. Improved and partially improved places can be purchased, the price depending upon the amount of improvement, fertility of soil, and nearness to transportation. Prices are rapidly enhancing as railroads are extended. It therefore behooves those who wish to settle in this county to come and make their selections soon.

Family Supplies and Furniture.

Provisions, groceries and all family supplies can be purchased at reasonable prices, being a slight advance upon first cost, with freight expenses added. Many of the stores are doing a heavy business and are located in nearly every settlement. Good family flour can be purchased at from $8 to $9 per barrel. Dry salted bacon 12 cents per pound. Butter three pounds for $1. Coffee four to five pounds for $1. Good sugar thirteen to fifteen pounds for $1. Florida syrup 60 to 75 cents per gallon. Dry goods, shoes, &c., at low prices. These are average prices in July, 1881. The intending immigrant need not bring heavy furniture, stoves, &c., as they can be purchased here. It may be well to bring bed clothes and wearing apparel, but feather beds may be left behind. All heavy and bulky articles had better be sold, as freights upon such articles are heavy.

Manufactures.

There are several establishments in the county that are manufacturing buggies, wagons, buck boards, carts, &c. Their work is substantial and suited to the character of our roads. There are two tin shops, one at Sanford and the other at Orlando, also harness and shoemakers.

Openings for Business.

There are openings for several steam saw mills, as the demand for lumber is greater than the mills can supply, and the rapid immigration which is now going on through the year, and the large number that will come in during the fall and winter, will rapidly increase the demand. The large number of boxes used for oranges and vegetables will warrant the putting up of a mill for the purpose of supplying the rapidly increasing demand. The material preferred for these boxes is pine, and timber that would not answer for building lumber could be used. There is needed a furniture factory with necessary machinery to make up our native woods and furnish goods here at reasonable prices, saving freight charges. Factories are needed to make up the cypress into barrels, tubs and buckets; also shingle machines; more starch mills to encourage the cultivation of cassava and arrow root, by creating a demand for the roots. Hides and deer skins are abundant, but no tanner to make them into leather and buck skins. There is a deposit of blue clay, suitable for the manufacture of pottery. A large quantity of jugs, jars, pans, &c., could be sold at fair prices. All these industries will pay. Water power is scarce, but fuel for steam engines is cheap and abundant. There are openings in various sections for good hotels, not only to entertain the large number of persons who are examining the country with a view of making it their home, but to accommodate the invalid and pleasure seeker.

Climate.

The climate of Orange County is a peculiar one, differing from countries north of us; mild in winter and not excessively hot in summer. Those who have not passed through these seasons here and tested the matter for themselves are apt to receive such a statement with doubt. Facts, however, prove contrary to what might be expected, that the summer weather is much more agreeable, and the heat less oppressive, than the same season in the north or west, and the nights are almost invariably cool enough to render some bed covering necessary before morning, enabling the sleeper to obtain refreshing rest, and even during the hottest days it is pleasant in the shade. The thermometer seldom gets higher than ninety-three degrees in the shade, and generally averages during the summer eighty-two to eighty-five degrees.

Those who have lived here for many years prefer the summers on account of the general uniformity of temperature, and the delightful sea breezes, which keep the atmosphere cool and pleasant. The weather does not debilitate, neither do people lose their energy; the rapid development taking place in this section is an attestation of the fact.

"The winter in Florida resembles very much that season which in the Middle States is termed "Indian Summer," except that in Florida the sky is perfectly clear and the atmosphere more dry and elastic. Rain but rarely falls during the winter months—three, four and not unfrequently five weeks of bright, clear and cloudless days occur continuously. This is one of the greatest charms of our winter climate, and forms a striking contrast with almost every other State in the Union," Mrs. Stowe writes. "The month of March has passed. Letters from our northern friends speak of its chilling blasts, its cutting winds, its long snow storms. Here in Florida it opened upon us in the perfume of orange blossoms, and we look on it now with the general remembrance of a long procession of sunny days, of blue skies and vivid green of blooming trees; of lettuce, radishes, and green peas in the garden; of roses and honeysuckles among the flower borders. It is true every day has not been equally bright and balmy. Changes of temperature here correspond to the severer ones of the North. Where they record a three days' snow storm, we remember a three days' rain storm, in which it has been about as chilly here as it usually is in a June rain storm at home."

The summer season is also peculiar. Showers are frequent, and in June and July commences what is termed the rainy season, which continues into September. There is rain nearly every day, but it seldom rains all day. The rain comes in heavy showers accompanied by thunder and lightning, and seldom lasts more than three hours. They generally occur in the afternoon and leave the remainder of the day with a bright sky and cool atmosphere. The soil being of a porous and absorbant character, the ground half an hour after a rain is dry and pleasant to walk upon.

Opinions are often formed of a country by comparison, sometimes resulting in a correct idea in other cases an erroneous one. Such is the case in regard to Florida. Thus when the line of latitude embraced by Orange County is traced across the Atlantic ocean, it takes in northern Africa and a portion of the great Desert of Sahara. Those who know anything of Florida are satisfied that no heated blasts such as pass over that land of desolation are known here. There are several reasons why there should be a difference: We are on a long narrow peninsula, passing through a number of degrees of latitude, having the Atlantic on the east and the Gulf of Mexico on the west, insuring a constant succession of sea breezes. Another reason is that a current of cold water from the north passes down our coast on the east between the land and the gulf stream, which does much to moderate the temperature. Located as we are upon the very borders of the torrid zone, we are relieved on the one hand from the rigors of a northern winter, and on the other from the extreme heat which is experienced in many of the southern as well as northern states during the summer.

The cold wave of December 30, 1880, passed over this county. The thermometer for a short time stood at twenty-eight degrees, but in a few hours rose above the freezing point; a thin skim of ice was formed. But little injury was done to vegetation. Tomatoes were killed; bananas and guavas were considerably injured, but have recovered, and pine apples, where entirely unprotected, were hurt some. The orange, lemon and lime were not effected, even the tender extremities of the limbs were not hurt, and the fruit upon the trees was not frozen or injured. In many places sweet potato vines, which are very tender, did not show the least effect of frost. Portions of the county that have large water protection on the northwest entirely escaped the effects of the cold. Another peculiarity of our climate is that when a cold spell does occur, it lasts but a short time.

The above lengthy discription of climate is necessary to properly understand why semi-tropical fruits and plants flourish in this section, and make their culture a success.

Health.

A very important question asked by those who are thinking of emigrating to Florida, and which demands a candid answer is, is your section healthy? The following facts must speak louder than any affirmation on our part. There are no acclimating fevers or diseases. There are no epidemic diseases as diphtheria, typhoid or typhus fevers. Yellow fever is confined almost exclusively to seaport towns, and only occasionally visits them. There has never been a case in our county, and there is no reason to believe there ever will be. Some portions of the county are low and swampy, with a rank and luxuriant vegetation, and where this is the case, the same types of disease will be found as elsewhere characterize such regions. Here such diseases are generally of a milder type than in other States, and yield more readily to proper treatment. A cogent reason has been assigned to account for this fact. The luxuriant vegetation which springs up in summer in other states passes through the putrifactive fermentation, this generates miasma; while in Florida (except during a portion of the rainy season) this decaying vegetable matter dries up before reaching the putrifactive state, and as a consequence the amount of malaria generated is much less than in climates more favorable to decomposition. The constant sea breezes tend to keep the atmosphere pure and carry off all noxious vapors.

Pernicious fevers and those diseases peculiar to semi tropical climates are rare occurrences; our malarial diseases assume the form of mild intermittants.

There are a large number of physicians in the county who have moved here to regain their health. Among them all there are only *seven or eight* who follow their profession for a livelihood. The United States census of 1870 gives the death rate for the State at large as 1.05 per cent., and this includes those cases of consumptives who came here too late to derive benefit from the climate. For further facts the reader is referred to the census of 1880.

Selecting a Residence.

In selecting a place for a residence, it is well to avoid the neighborhood of swamps, hammocks, dark water lakes and sluggish streams. Experience has proved that a residence only a mile, and in some instances a less distance from unhealthy places, especially if a body of timber intervenes, renders the location healthy. The high pine lands are noted for health. It is safe to move here during any portion of the year, either winter or summer. A majority of people, however, prefer to come in October or even later.

Orange County as a Home for Invalids.

In the cure of diseases, both pulmonary and bronchial, physicians unite in the opinion that patients should reside during the winter months in a dry climate of uniform temperature, where they can take daily exercise in the open air and enjoy the benefit of sunshine; and that nothing can be more detrimental than confinement in an artificially heated atmosphere. Another prerequisite insisted upon by many writers upon these diseases, is that the residence should be in a section free from germs and gaseous products of decomposition. That the test of atmospheric purity is the presence of large quantities of *ozone*. Observations made by the United States Signal Service, at Denver, Colorado, show that the mean monthly amount for six months is but 3.8, while for the same months and same length of time the observations made at Jacksonville, Florida, at the Signal Service office, indicate 6.2, nearly twice as much as Colorado. Another requisition is that the air should be *dry*. A series of observations continued for five years show that Florida possesses a less mean relative humidity than the State of Minnesota. These facts indicate that we have a dry and pure atmosphere with a mild winter climate, conjoined with opportunities of fishing and hunting as recreations, making this section pre-eminently attractive as a home for the invalid. Dr. Lente, writing upon our climate, says: "In Florida the sun shines so brightly, the air is so balmy, the song of the birds so enlivening, and the orange trees in their bloom or ladened with their golden fruit, lend such a charm to the outlook from the windows, that the most indolent or the most cold-blooded invalid feel little inclined to stay indoors.

An eminent northern physician writing in regard to the fitness of Florida, more particularly as a place of resort for invalids, and especially for consumptives, says: "She is our southern Italy, but more favored by geographical position than Italy, since she lies almost in the very fountain of that gulf stream whose waters are freighted with a genial atmosphere; and also because her winter climate is comparatively free from depressing humidity, which debilitates the healthy and exhausts the invalid both in Italy and Cuba. This is a point of such vital importance that you must bear with me while I insist upon it. To find a climate which shall at once be mild and dry, is the first requisite for all who are suffering under chronic rheumatism and pulmonary affections. On this point there can be no disagreement among physicians unless it be on the exact degree of temperature; and those who have personally tried both the cold, and the dry warm climate, will very generally yield to the greater attractions which keep one out of doors in the warm climate. The theory of the cold dry air may be right for those who can resist the cold, but the practice, nine times out of ten, is wrong. Let it be understood, once for all, that if the consumptive is to live at all, he must be in the open air. Does the peninsula of Florida offer this desirable climate? I unhesitatingly say yes. One comes here in mid-winter from the north, where he had been imprisoned by the pitiless cold, and straightway begins to breathe and live again. You will pardon something to the sunny influences that fill him with enthusiasm. Besides I find here hundreds of hopeful health-seekers full of joy at their liberation, and many who came years ago as a last resort and who now give every evidence of restored health."

Unfortunately many put off the journey to our land of sunshine and flowers until the fell destroyer has taken such a hold upon them that they must die. Better for such to remain at home and pass their last hours among friends. It is sad to see one die among strangers uncheered by the presence of loved ones.

There are many in this county who are thankful that they were induced to come to Florida in time to be benefited, and have remained here. The writer is among this number, and can attest, by renewed health, that the climate possesses a healing and recuperative influence. The country offers a refuge to those suffering from pulmonary and bronchial diseases; to those afflicted with asthma, catarrh and rheumatism. To derive benefit from the climate the visit should not be delayed; come while you are yet able to take exercise by walking in the open air; come, if possible, when the first symptoms of disease manifest themselves; come while there is hope of regaining such a degree of health as will enable you to pass the remainder of your days in comparative comfort.

Facilities of Transportation.

To make the growing of semi-tropical fruits and early vegetables a success requires both rapid and reliable transportation facilities. The following statement of the facilities possessed by Orange County to get products to market, shows that they are equal, and perhaps superior, to any other county in South Florida: The great outlet of the county is the St. Johns river, flowing north and emptying into the Atlantic. This gives us access to the port of Jacksonville, with its sea-going steamers and the various railroads which center at that city. Navigating this river are elegant side-wheel steamers, the number of which is being added to each year, to accommodate the increasing travel and traffic. There has been during the past season over twenty arrivals of steamboats weekly at Sanford on Lake Monroe.

The Wekiva river, taking its rise near the center of the county, is navigable for a distance of twenty-five miles, and enters the St. Johns north of Lake Monroe. A large portion of the traffic of this section of the county passes over this river. The shipping point is Clay Spring, distant from Apopka four miles. The semi-tropical scenery on this river is very fine.

The Ocklawaha river on the west has its source in Lake Ahapopka, passes through a chain of large lakes, and enters the St. Johns opposite Walaka, and gives a water transportation for that portion of the county. A number of steamboats run on this river from Jacksonville and Palatka.

A company with ample means are now engaged with dredging machines in deepening the channel of the Ocklawaha where it passes out of Ahapopka into Lake Dora, and when completed will give transportation facilities by this route to those living on and near the great lake, besides draining many thousand acres of very rich land.

Railroads.

There are two railroads now in operation in the county. One, the St. Johns and Lake Eustis railroad, starts from Astor on the St. Johns, just south of Lake George, and passes in a southwesterly direction to Fort Mason, on Lake Eustis. This road furnishes more rapid transit than the steamers on the Ocklawaha, and is a good route by which to reach the interior of the county. The amended charter authorizes the company to extend their road by way of Apopka to Orlando, the county seat, and will pass through a thickly settled portion which is rapidly developing.

Another road is in operation from Sanford, on Lake Monroe, to Orlando, and opens up the central part of the county. This company proposes to commence, in a short time, the construction of a road from Longwood (a depot on their road) to Apopka, and make it the main line to Charlotte harbor, on the Gulf. The citizens along the route of the proposed branch have contributed toward its construction the sum of thirty-eight thousand seven hundred dollars. This road will give transportion facilities to South Apopka, where are some of the richest lands in the county, with superior water protection on the northwest. This company is now extending the road from Orlando south to the Kissimmee river, and also contemplate a branch to Titusville, on Indian river.

Proposed Railroads.

There are quite a number of proposed railroads, which have already obtained charters: one from Lake Jessup to Orlando, partially completed, has been sold to heavy capitalists having large landed interests in the county, who propose to extend it by way of Apopka to Fort Mason, thence to Ocala and connect with the roads uniting there. Another from Sanford by way of Fort Reed to Lake Jessup will pass through the oldest settlements in the county, with large groves and good lands, and is to be built entirely with local capital. It is estimated that one crop of oranges from the groves along the proposed line of road will build it.

A road has been chartered to run from Clay Spring by way of Apopka to Lake Ahapopka, and will be a little over eight miles in length to the lake, upon which will be placed a steamboat, bringing this popular section in direct communication with the St. Johns. There is an important line being surveyed from Sanford by way of Sorrento to Lake Eustis and Fort Mason. It is also the intention to construct a branch from Sorrento to Apopka.

Thus it will be seen that there is no part of Orange County that will not be within ten or twelve miles of either navigable water or a railroad. Taking this in connection with its geographical position and climate, Orange County becomes *par excellence* the region for fruit growing and vegetable raising.

Development.

No one but an old settler can have a just conception of the rapid changes that have taken place during the last twelve years, and more especially during the past six years. At that period the great extent of territory embraced in Orange County was an almost unbroken wilderness, with the exception of a few houses at Fort Reed, and but one house at Mellonville, on Lake Monroe, used as a store, and a small village at Orlando, the county seat; also some plantations on South Apopka. The remainder was a primeval forest, with an occasional clearing, the residence of cattle men; the principal business was cattle raising. Nearly all else was a vast solitude, the home of deer, panther and bear. One steamboat arrival weekly at Mellonville was sufficient for both passenger and traffic business, not only for this county, but for points south. There were but three post offices in the county, and neighbors living at a distance went by turns to the nearest office for mail. There was but one Masonic Lodge south of Palatka (except at Key West), and was located at what is known as Apopka. The population in 1867 was 1,516; in 1880 about 7,000. This may suffice as a brief description of the past; the present is a bright contrast. Now the cattle men's cabins have given way to the neat residences of the immigrants, surrounded, in almost every instance, with trees and plants of semi-tropic growth. Orange groves are seen in nearly every place adapted to their growth.

Post offices have increased to thirty-seven, and these scarcely supply the demands of the numerous settlements; and there are four money order offices, being a greater number than any other county in the State. Three telegraphic offices keep us in communication with other portions of the country. Masonic Lodges have increased, there being one at Sanford, Orlando, Apopka and Fort Mason, and over twenty weekly arrivals of steamboats at Sanford are scarcely able to supply sufficient transportation.

There are four newspapers published in the county, one at each of the following places: Sanford, Orlando, Apopka and Lake Eustis; they are well patronized. There are forty-five public schools, furnishing educational facilities for all the children. There are about the same number of churches, belonging to Presbyterian, Methodist, Baptist, Episcopalian and Lutherens; a number of the buildings are neat and commodious. Towns and villages have sprung up rapidly in various sections; the most prominent are Sanford, on Lake Monroe, and Fort Reed, three miles distant. Maitland, Orlando and Apopka, in the interior, and Lake Eustis and Fort Mason on Lake Eustis. There are thriving settlements at Sorrento, Tangerine, Zellwood, South Apopka, Lake Conway, Altamonte and Lake Jessup, beside a number of others.

All this in connection with railroads and increase of population, is sufficient to indicate that the development of the past has been rapid, and the lookout for the future is very bright.

Why Settlers Should Select Orange County.

The climate enables a person to work out of doors throughout the whole year, winter and summer alike, and the cool nights make sleep refreshing. Tight, expensive houses are not required, and heavy clothing is not needed. The farm work, which at the north must necessarily be performed within the limits of about six months, may here be allowed the whole year. Thus the farmer is not hurried by the shortness of the season, and be constrained to overwork himself. Corn is made and ready to be housed by the first of August.

The time and expense required to obtain an abundant supply of fuel is saved, as but little more is required than is used for cooking purposes. During a large portion of the winter months doors and windows are left open, unless the weather is exceptionally cold.

Nearly every portion of the county is within easy reach of transportation, either by water or railroad, and these facilities are being rapidly extended. (See article, Facilities of Transportation.)

The county being generally healthy, doctor's bills are saved.

Churches of different denominations and public schools are found in every settlement.

Though the weather in summer is warm, there is an exemption

from sunstroke, and dogs do not go mad; there are no cases of hydrophobia.

The people are moral and law-abiding, and cordially welcome all, irrespective of political faith, who come to make this their home and grow up with the development.

This section is the great sanitarium, where the invalid may reasonably expect to derive benefit from a mild and uniform climate.

The population is composed of people from all sections of the country and from Europe, members of different political parties and of different religious faith, who get along harmoniously.

This is a *white man's* country: there are but few negroes among us.

All the products of the county find a ready sale at remunerative prices. Fruits and early vegetables find a market at the North and West. The great variety of fruits, vegetables and agricultural products that can be raised with profit, gives the settler an opportunity of selecting the particular branch of culture that may suit his inclination, and make it a specialty.

There is another important fact that should not be overlooked in selecting a home: that Orange County is south of the line of injurious frosts, which makes the raising of early vegetables a business that can be depended upon, as well as enabling the farmer to cultivate the more tender semi-tropical fruit. During the past winter in many places in the county, sweet potato vines, which are very tender, did not show the least effect of cold.

Addenda.

There are many fruits and plants that have not been in cultivation a sufficient length of time to determine whether they are fully adapted to this latitude, or sufficient attention has not been given to positively assert that they will prove valuable additions to the already large list of products, therefore they have not been mentioned in the body of the pamphlet, preferring to state what has been accomplished, and which may be repeated by others.

But few who have not had their attention directed to the subject can realize the value of fibrous plants as farm crops, or the immense quantities required to supply the demand of manufacturers, a large proportion of which is imported, and that by raising them at home millions of dollars would be saved to the country.

Sisal hemp will grow upon land worthless for any other purpose. A ton of cleaned hemp can be made to the acre, worth $300 per ton. In addition to this plant, there was introduced from Yucatan the pulqua and centuary plants, which are the main dependance of that country for material for cordage and rope, of which they export large quantities. They have become perfectly acclimated in Florida.

Ramie fiber, which is so extensively used in dress goods, only awaits the invention of machinery to successfully clean it at small cost. Large quantities are now being raised in Louisiana, and there is no reason why its cultivation should not be remunerative in this State. East India jute has proved from the experiments made to be well adapted to our climate, and can be planted upon land too wet for other crops. The result of the experiments made in this county show that from 3,000 to 4,000 pounds of fiber can be made upon an acre.

Spanish cockle burr or ceaser weed, now called Florida hemp, is found growing wild; takes possession of fence corners and is hard to get rid of. The fibre is strong, intermediate between flax and hemp. The attention of manufacturers has lately been directed to this plant, which must prove of great value. It is easily raised, and with cultivation will grow to six or eight feet high.

Scrub and saw palmetto, heretofore considered a pest and worthless, has now a commercial value for the purpose of making paper pulp. The supply of the raw material is inexhaustible, and can be supplied in any quantity.

Grapes, Fruits, Nuts.

A large variety of grapes from the north and west have been tried, and some of them have proved to be adapted to this section; such as the Delaware, Maderia, Hartford and Dina. The Scuppernong is particularly a southern grape; adapts itself to almost every location, and is free from the attacks of insects; produces an abundance of fruit, and makes a good wine. The Catawba ripens unequally, rots, and cannot be depended upon. I have had two crops ripen the same season, and a third crop set.

The Surinam cherry does well, and the papaw, tamarind, mango and camphor tree are doing well when planted in sheltered places on the south of large bodies of water. Pecans grow well upon any soil that will produce hickory. A sufficient number of persons have planted tea to demonstrate the fact that each family can raise sufficient tea for family use, and that it is of superior quality.

The Olive has been extensively distributed by the Commissioner of Agriculture, and promises to be a success. The tree fruits in nine years from the seed, and is easily propagated from cuttings. The castor-oil plant grows to the size of a tree, often lives for years, and continues green during the winter.

Each year new plants and fruits are being introduced, and such as prove to be adapted to the country, add to the already large resources and material wealth.

MAITLAND, Orange Co., Fla., Aug. 16, 1881.

Dr. Z. H. MASON, Apopka, Fla.,

Dear Sir:—Most gladly do I respond to your request for a statement of the beneficial effects of the climate upon me. I came to Orange county about seven months ago from Chicago, where for several years I had suffered much from catarrh; for years my headaches, lasting for days in succession, were almost unendurable. Now, I have hardly a vestige of the disgusting disease left with me; have no headache worth mentioning, and consider myself *cured.*

I am enthusiastic over the climate, and shall do all I can to let the thousands in the north who are affected as I was, know what a residence here will do for them. Very truly yours,

LORING A. CHASE.

How to Get Here.

From New York, Philadelphia and Baltimore there are regular steamers to Jacksonville, Fernandina and Savannah.

Regular steamer rates from New York or Philadelphia to Jacksonville are, including meals, First Class, $25 ; Emigrant, $13.

From all northern cities there are through routes by railroad to Jacksonville and other points in Florida, with parlor and sleeping cars and fast freight lines.

Present regular rates by railroad are about as follows :

			First class.	Second class.
From Boston to	Jacksonville		$37 50	$30 00
" New York to	"		31 00	25 50
" Philadelphia to	"		29 50	22 00
" Baltimore to	"		28 00	20 50
" Richmond to	"		26 25	20 00
" Cincinnati to	"		25 60	22 50
" Chicago to	"		34 85	28 85
" St. Louis to	"		32 20	26 60
" St. Paul to	"		48 85	39 85

The time from New York to Jacksonville by rail is now reduced to forty-six hours. From Philadelphia to Jacksonville, forty-three hours.

At Jacksonville the traveler connects with steamers on the St. Johns river, which land at any of the river points in the county.

At Astoria and Sanford connections are made with lines of railroads running into the interior of the county.

ORANGE COUNTY has made greater progress in the successful cultivation of the orange and other fruits, the product of a genial climate, than any
county in Florida. The successful orange culture there, and its large pecu-
niary returns, are assured. ZELLWOOD is situated near the west-central part
of Orange County, upon the watershed of the Florida peninsula, latitude
about 28° 3′ N. To reach ZELLWOOD, the most direct route is by steamboat
from Jacksonville to Astor, the terminus of the St. John's and Lake Eustis
Railroad. The other terminus of this road, at Fort Mason, ends for the
present within 12 miles of ZELLWOOD. Near this terminus, at Lake Eustis,
carriages may be procured for conveyance further. Another, and a very
interesting route, is by the Ocklawaha River to Clifford's or Badger's wharf,
on Lake Eustis, 12 miles from ZELLWOOD. That portion of Orange County
near ZELLWOOD is studded with very beautiful clear-water lakes, the surface
of the country is high and rolling, timbered with large pine trees, and re-
markable for its healthfulness.

MAP OF
ORANGE CO,
FLORIDA,
Showing ZELLWOOD
& Vicinity.
1881.

Completed R.R.
Proposed " "

For further information, address

T. ELLWOOD ZELL,
1111 Arch Street, Philadelphia.

Or, J. A. WILLIAMSON,
Zellwood, Orange County, Florida.

Arrangements will be made for the clearing of land, planting of groves, and their subsequent attention.

Address,

J. A. WILLIAMSON,
Zellwood, Orange County, Florida.

THOMAS EMMET WILSON,

ATTORNEY and COUNSELLOR AT LAW

Postoffice - - - - - - - Sanford, Fla.

Residence - - - - - - Sylvan Lake.

H. B. LORD.

WATCHMAKER and JEWELER.

---DEALER IN---

WATCHES, CLOCKS, JEWELRY and FLORIDA CURIOSITIES.

SANFORD. · · · · FLORIDA.

WM. EMERSON.

Carpenter and Builder

SANFORD, FLORIDA.

Will contract for and build Residences, Bridges, etc., in any part of Orange or Volusia Counties. Will also do a general carpenter jobbing business, and warrant satisfaction. Counters, Shelving, Desks, and Inside Finishing a specialty. Also buildings moved and raised.

SIRRINE'S BOARDING HOUSE,

SANFORD, FLORIDA,

Northwest Corner Fourth Street and Palmetto Avenue,

is now open for guests.

HOTEL FARE AT BOARDING HOUSE RATES.

Three minutes walk from Railroad and Steamboat Landing.

WM. SIRRINE, Proprietor.

FLORIDA LAND AND IMPROVEMENT
COMPANY.
(HAMILTON DISSTON'S PURCHASE)

4,000,000 ACRES OF LAND

'Orange County Colony'
OF FLORIDA.

50,000 ACRES IN ORANGE CO.

Near Full-Grown Orange Groves and Beautiful Lakes.

NEW YORK OFFICE, 115 BROADWAY, ROOMS 111 and 113.

The Florida Land and Improvement Co., of New York, having purchased from the State of Florida, Four Million Acres of Land, intend placing the same in the market for sale at low prices, and upon easy terms.

The first Colony will be located in Orange County, near Orlando, the county seat, and in the vicinity of Lake Conway—upon 20, 40 and 80 acre farms—at prices ranging from one dollar and twenty-five cents per acre, to five dollars per acre.

The quality of the soil, the equable climate, and the vast productions of the State need no exaggerated statements to induce any person to locate in any of the Southern Counties of Florida.

The climate is very similar to that of California, and a farmer can employ his time every day in the year, without fear of extreme heat or cold.

We are now preparing for publication, a pamphlet on the "Resources of Florida," which will be forwarded by mail, upon receipt of stamp.

Colonies desiring large tracts will obtain special rates and long credits

WM. H. MARTIN, Land Commissioner,
Florida Land & Improvement Co., 115 Broadway, N. Y., Rooms 111 & 113.

Agent in Chicago, Ill., W. H. NICHOLS, 101 Clark Street.
Agent in Buffalo, N. Y., J. T. McLAUGHLIN, 350 Main Street.
Agent in Jacksonville, Fla., COL. I. CORYELL.
Agent in Orlando, Fla., MAJ. M. R. MARKS.

FOR GOOD BOARD,

BY THE

DAY OR WEEK,

GO TO

G. E. SAWYER'S,

SANFORD, ORANGE CO., FLA.

JOHN DODD,

DEALER IN

Dry Goods, Groceries,

NOTIONS,

Boots and Shoes, Hats and Caps,

CLOTHING.

QUEENSWARE, WOODENWARE, TOBACCO, CIGARS, &c.

SANFORD, FLORIDA.

CHARLESTON LINE.

———o———

Steamships leave

NEW YORK

WEDNESDAYS and SATURDAYS,

Connect at

CHARLESTON

with iron palace steamer ST. JOHNS for

FERNANDINA, JACKSONVILLE,

and all points on St. Johns, Ocklawaha and Indian rivers.

JAMES ADGER & CO., Agents, New York and Charleston S. S. Co.,
Charleston, S. C.

W. A. COURTENAY, Agent, New York and South Carolina S. S. Co.,
Charleston, S. C.

RAVENEL & CO., Agents, C. and F. Steam Packet Company,
Charleston, S. C.

R. J. ADAMS, Agent,   H. GAILLARD, Agent,
Palatka, Fla.   Jacksonville, Fla.

J. L. HOWARD, Contracting Agent, Jacksonville, Fla.

Sinclair's Real Estate
AGENCY.

If you want an Orange Grove anywhere from $300 to $20,000 in value;

If you want unimproved land convenient to transportation, from one acre to thousands;

If you want to purchase, or procure reliable information in regard land or prospective railroads;

If you want first-class High Pine Land;

If you want Hammock Land;

If you want Bay Land;

If you want Saw Mills or Hotel Sites near sulphur springs;

If you wish for reliable information in regard to climate, products of soil, expense of clearing land, cost of trees, expense of setting out, and caring for Groves;

If you desire to be put in correspondence with successful farmers and orange growers;

If you desire information in regard to best routes to reach Florida from any section of the country, either for freight or passengers;

Address **JOHN G. SINCLAIR,**
No. 1, DeLaney's Block, Orlando, Orange County, Florida.

If you want an

ORANGE GROVE,

or a lot of wild land to plant one on,

Below Damaging Frost Line,

South of Lake Monroe, the head of *certain* navigation on the St. Johns river, in that portion of the peninsula where the lands are generally

TOO POOR TO BE UNHEALTHY

where you can obtain

GOOD SOCIETY, GOOD SCHOOLS, GOOD CHURCHES. GOOD WATER. GOOD FEELING. GOOD TRANS-PORTATION AND FEW INSECTS,

where almost every man you meet is a *Non-commissioned Land Agent*, who wants to speculate on you. come to my house at this place, and

I WILL SHOW YOU WHAT YOU WANT,

Judging entirely by the amount of cash you want to invest, and I will buy you the land as cheap as it can be had. I am handling over

2,000,000 ACRES OF LAND,

from $1.25 to $250.00 per acre, and over

100 Orange Groves,

from $1,000 to $75,000. I will deal with you justly and honorably. I have settled over 900 good men in Orange, Volusia and Polk counties.

I represent myself in Orange, John Snoddy of Bartow in Polk, and E. E. Ropes at Volusia in Volusia.   **M. R. MARKS**

Real Estate and Land Agent.

WILCOX, Orange County, Fla.

NOT A CHEAP BOARDING HOUSE.

Those in search of a delightful winter home cannot fail to be pleased with

ZUFRIEDEN,

A beautiful winter resort, situated in a high, healthy country, on

LAKE CHARM,

One and a half miles from

Oviedo, Orange County, Florida,

And two miles south of Solary's Wharf,

LAKE JESUP.

Steamers from Jacksonville and Sanford make regular landings at the last named place, and parties coming to, or returning from Zufrieden, will have but two miles to ride, over a pleasant road, through natural forests, and past orange groves covering hundreds of acres.

THE COUNTRY ON THE SOUTH SIDE OF LAKE JESUP IS HIGH AND HEALTHY,

and the soil, mostly GRAY AND BLACK HAMMOCK, IS THE BEST KNOWN FOR EARLY VEGETABLES AND TROPICAL FRUITS. THE ORANGE GROVES IN THIS VICINITY ARE THE FINEST OF THEIR AGE IN THE STATE OF FLORIDA. These are facts of which you will be fully convinced on making a visit to this favorite section and investigating for yourself. No orange trees have ever been injured by frost or cold south of Lake Jesup. The people are intelligent and enterprising, and the country is improving with great rapidity. Methodist, Baptist and Temperance societies hold regular meetings.

FOSTER'S CHAPEL,

A commodious and beautiful church, named after Dr. Wm. Foster, of Clifton Springs, New York, at whose expense the building was erected, and whose winter residence is on Lake Charm, is in the vicinity. Our house is well finished and furnished, and the table is supplied with fresh vegetables, fish, game and tropical fruits.

TERMS:

From fifteen dollars to twenty-five dollars per week, according to location of rooms ; or from two dollars and fifty cents to four dollars per day.

For further particulars address   MRS. J. L. BREWSTER,
Zufrieden, Lake Jesup, Florida.

The town of Tavares is located on a high ridge, on a peninsula formed by Lake Eustis and Dora and the Ocklawaha river, and embraces nearly one thousand acres of oak, pine and high hammock land, all of which rank as first-class orange land. A large orange grove, just north of the town, on Lake Eustis, now in full bearing, the property of the proprieters, Messrs. St. Clair-Abrams & Summerlin, attest the quality of the land and its adaptability for successful orange culture.

The town is laid out in lots suitable in size for business purposes, villa sites, hotels and manufactories. The streets and avenues vary in width from seventy to one hundred feet. The principal avenue, called Tavares Boulevard, extends along the banks of Lake Dora over a half mile and varies from one hundred to one hundred and fifty feet in width. On this avenue the proprieters will expend a large sum of money, to make it one of the most beautiful and attractive drives in the State. On the eastern end of the Boulevard, lots have been set apart upon which to erect winter hotels. On one of these lots, fronting Shore Park, will be erected next year the largest hotel in the State. The hotel will contain one hundred and fifty bed-rooms, beside other rooms necessary for a house of this size.

The view from the town across Lake Eustis on the north and Lake Dora on the south is unequalled for beauty in the State. The Ocklawaha river on the west, uniting the two lakes, is a tributary of the St. Johns and is noted for its beautiful scenery.

Three railroads will center at the town. Work on the Tavares and Lake Monroe railroad has been commenced and its early completion to Sanford is a fixed fact.

A hotel with accommodations for one hundred guests, and thirty cottages are being built. The largest lumber mills in South Florida are located at Tavares, the capacity of the mills being twenty thousand feet of lumber per day. Calcium lights have been ordered, to furnish the necessary light to enable the mills to run night as well as day, in order to keep up with the demand for lumber.

The present transportation facilities are unsurpassed by any town in the State. Steamers from Jacksonville via Ocklawaha river land on the wharves of Lake Eustis the year round. From Fort Mason, the present terminus of the St. Johns & Lake Eustis Railroad, distant four miles, steamers arrive daily.

The healthfulness of Tavares is noted; the easterly winds which prevail during the summer months blowing across open piney woods for nearly twenty miles.

Church and school privileges rank among the best in the State. Arrangements have been made for the immediate erection of a school building, and lots have been set apart on which to build a church for every denomination, to aid in the building of which the proprietors will donate lumber sufficient to erect a suitable church edifice on each lot, so that the expense to the congregation will be very light.

The lumber mills, the orange growing, and market gardens, the erection of buildings and the construction of railroads furnish employment for a large number of skilled and unskilled laborers, and the demand is increasing. Good opportunities are open for all the different branches of business and one hundred men now find employment in Tavares.

The proprietors own and keep constantly employed a steamboat, (in addition to the other boats plying on the lakes,) in connection with their lumber mills. The lumber mills at Tavares are the only ones furnishing orange boxes and vegetable crates in a territory now producing seventy thousand boxes of oranges and a like number of crates of vegetables per annum, and the demand will increase rapidly in the near future. The capacity of the mills is fully equal to the requirements at present and will be kept up in the future.

In addition to the town property, the proprietors own a large tract of first-class orange and vegetables lands, situated from one mile to three miles of the town, which will be sold to actual settlers for improvement at $5.00 per acre and upward.

Lots will be sold at reasonable prices to actual settlers and facilities given for improving homes.

For further particulars, prices of lots, etc., address

St. Clair-Abrams & Summerlin,

Tavares, Orange Co., Fla.

APOPKA CITY.

Apopka City is located in the center of the county, upon high rolling land, at a distance from swamps or local causes of disease, having been selected in 1853 as a residence, by a physician, on account of its dry atmosphere, elevation and good water. It has between three and four hundred inhabitants, and each residence surrounded by orange trees, and consequently covers a large area of ground.

The public buildings are a Town Hall, Masonic Lodge, neat Methodist and Baptist Churches, with a prospect of soon having a Presbyterian Church edifice. A Drug store and four stores doing a general mercantile business, with fair stocks of goods, which are sold at reasonable prices. A Post Office and Money Order Office, with a Daily mail from Sandford and Orlando, and a Semi-weekly mail from Fort Mason.

The Public School has an average attendance of one hundred pupils. The South Florida Citizen, a weekly paper is published here. A Musical Academy to be under the charge of Professor John Esputa is now being erected. We have a Wagon and Blacksmith shop, and a Steam Saw-mill, with the prospect of another shortly. Also a regular Beef market.

The facilities for transportation are rapidly increasing. An extension of the South Florida Railroad from Longwood through Apopka City to Charlotte Harbor on the Gulf of Mexico, will in all probability be finished to this place by the 1st of January. The citizens on this portion of the route, have contributed $38,700 towards its construction.

There is another proposed line from Orlando by way of this place to Fort Mason. By way of Clay Spring, four miles distant, we have water transportation to the St. Johns, twenty-five miles and connection with steamers on that river. A company is now engaged in deepening the channel of the Ocklawaha river, where it passes from Lake Apopka into Lake Dora, distance from Apopka to landing on the lake three and a half miles ; this will give water communication with Fort Mason and the railroad at that place.

There is a fine opening for a good hotel, which would be well patronized. There are no drinking saloons here, and the citizens are determined to use all lawful means to keep them away.

The Lake Eustis Region.

————o————

This beautiful region is situated in the northern portion of west Orange, and embraces the scope of country around the western, northern and eastern sides of Lake Eustis. That on the west extending from the Ocklawaha river (forming the boundary line of Sumter County) to Lake Yale ; the northern section consists of the Fort Mason neighborhood and includes the country along the St. Johns and Lake Eustis Railroad : and the eastern part, the territory between Lakes Eustis and Dora, and extending to the Ocklawaha river joining these two lakes.

Nearly the whole of the country indicated consists of fine, high, rolling pine lands, much of it first-class South Florida land, here and there hammock lands are found in small areas, and there is but a comparatively small portion that can be considered worthless, or composed of scrub, grass-ponds, marshes, or swamps ; but consists of a yellow sandy soil covered with a growth of large pines and interspersed with beautiful, clear, high bank lakes ; and admirably adapted to semi-tropical fruit culture, as is evidenced from the number of flourishing bearing orange groves seen in various places, and the different fruits thriftily maturing in the different seasons of the year, and the great number of fine, fast growing, large, young orange groves already started, and the numerous others being under way.

The lands within our region have been almost entirely taken possession of by homesteaders, nearly all of whom have lands for sale, and good bargains can be obtained by timely prospecting and application, Very many excellent locations to suit all tastes can be secured at very reasonable rates.

The general description of Orange County is applicable to this region in every particular in all meritorious attributes, as we enjoy all the advantageous characteristics possessed by any of the other sections of our county, and furthermore claim a preeminence in some respects, especially as regards our unique and fortunate geographical location, which will be presently adverted to. The three great requisites which should always engage the consideration of the prospective settler as matters of the first importance, we can with pleasure confidently present. These are, a country of extreme healthfulness, convenient transportation, by water and rail—the Ocklawaha river and the St. Johns and Lake Eustis Railroad—and a fairly good soil.

The Lake Eustis region also offers unusual inducements to those desirous of engaging in winter vegetable gardening ; our southern situation and our soil combined, the many localities here having good water protection against frost and the direct transportation render quite propitious the conditions for the prosecution of this business. It has been engaged in here somewhat extensively and quite successfully and lucratively for several seasons past and is destined to become a resource of great commercial importance. And here it is well to observe that besides the almost incomparable wholesomeness of our highland country and the general beauty of its landscapes and magnificent lake views, it

presents many attractions and charming merits for invalids, winter
visitors, tourists and sportsmen.

Lake Eustis being one of five large lakes (of the Great Lake Region)
forming the head-waters of the famous Ocklawaha river, and lying in
close proximity to Lakes Dora and Harris, two of the four others, possesses from its situation peculiar advantages that do not belong to the
other lakes and other parts of Orange County; among others, it may be
mentioned that it is the *Key* so to speak, as regards transportation
facilities and improvements through a large scope of the surrounding
country. The improvements in the opening up of the navigation of the
river and Lakes Dora and Apopka above, now being prosecuted, will
redound to the benefit of this Lake Eustis region. And there is a great
likelihood of railroads being made on *all* sides of this lake. There are
already six or seven roads recently chartered, and a great probability of
one half or more of them being made. Besides the St. Johns and Lake
Eustis Railroad already in operation, whose terminus is at Fort Mason,
(on the lake,) two or more railroads to the north and the westward which
are to be the grand outlets and thoroughfares of our great county to directly connect her with the principal markets of the world, and the grand
systems of northern railroads, *must necessarily* pass through our region
and close by our lake.

So we are situated at the *gateway* of commerce of our county and
an extensive region adjacent, for here will be the traversing and crossing
of such a number of roads as to eventually make it a center of no small
importance. And furthermore, the most important railroad lines connecting the city of Jacksonville most directly with South Florida, either
to the Atlantic, or Gulf Coast most advantageously, most certainly must
traverse our locality.

So it can at once be seen that the immigrant can not well go amiss
in casting his lot among us, and selecting a home hereabouts. And in
our endeavors to engage attention and enlist an interest in our community, it is with a pleasing consolation and pride that we can inform the
new-comer that this region is inhabited by excellent society and posses-
es good facilities in the way of school and church privileges, con-
venience to well stocked stores with goods at reasonable rates, also to
hotels, saw-mills and workshops, &c., to be found at the flourishing
villages of Fort Mason, Eustis and Tavares. We have a telegraph office
at Eustis and a weekly newspaper, the Semi-Tropical is also published
there. And withal, the prospects are quite favorable for this being
the central part of a new county, to be established in all probability, at
no distant day.

Few localities in Florida are receiving greater accessions to its
population, or are improving faster in every respect than this, and we
conscienciously assert the belief that none will make more gigantic
progressive strides in the near future. We already possess very nearly
if not all the desirable prerequisites for the creation of attractive and
the most pleasant of homes, the rendering of life delightful and happy
to its utmost with such Eden-like surroundings, with the fewest of its
ills, and therefore all that is conducive to the maintenance of a healthy
status and the fostering of longevity.

Much might be said about the beauties of our climate, our multifarious productions, our various resources, &c., but space forbids further expatiation, and description, and as "seeing is believing," we cordially invite all home seekers and health seekers to come and see us and our fine country.

Those at a distance desirous of further information can address Hon. J. M. Bryan, E. G. Rohrer, R. McS. Byrne and Dr. E. B. Miles at Fort Mason, or J. A. McDonald, Col. G. H. Norton and A. St. Clair Abrams, Lake Eustis P. O., Orange Co., Fla.

ROYALLIEU.

All those contemplating growing oranges or lemons should go to Royallieu. Royallieu is situated among the high rolling hills, among the lakes in West Orange, with Post Office, Telegraph and Public School facilities, with public roads running from Sanford to Lake Eustis, and from Orlando to Lake Eustis.

For further information address

Tangerine and It's Surroundings.

In looking on a good map of the State of Florida, near the head of the Ockiawaha river, will be found Lake Dora: Lake Beauclair lies just south of it and, with the river as it passes northward from Lake Apopka, form the western boundary of our locality. We are located midway between Royallieu and Zellwood postoffices. We are ten miles from the present terminus of the St. Johns and Lake Eustis railroad, at Fort Mason. In connection with the railroad, a little steamer plies about Lake Eustis. A small expenditure in improving the channel between Lakes Eustis and Dora would enable this steamer to come to the various proposed landings on Lake Beauclair and the Ocklawaha river along the western border of our region. This, it is hoped, will be accomplished within a year.

The region which we are trying to bring more thoroughly to the attention of persons seeking first-class orange-growing, healthful locations includes twenty sections, mostly of high, rolling, first-class pine land. Our locality is peculiarly adapted to semi-tropical fruit culture, on account of its freedom from killing frosts, the effect of lying to the east and southeast of Lakes Beauclair, Dora, Eustis, Harris and Griffin.

Improved and unimproved land can be had at reasonable prices in lots to suit purchasers, from private parties. Our neighborhood is rapidly settling up about ten new families having come in within the past year. Our population (all white save one family) is of the quiet, orderly class, and we have no liquor saloons within ten miles. Our locality is devoid of malaria. No doctors needed here, if they propose to obtain a living by their profession.

The lakes of this region are well stocked with fish—the varieties being trout (black bass of the north) cat and bream. Deer are found on the hills, especially in some "scrub" or waste land adjoining us, and in the "flatwoods" near by. Quail and squirrel are also found, and doves and blackbirds sometimes become a nuisance. Sunday school is held each week at Zellwood, and also near by at Sorrento. Episcopal church service is held monthly at Zellwood, and four miles south of Tangerine the Methodists hold meetings once or twice monthly. A public day school will certainly be established within our borders during the coming fall. At the proposed town of Tangerine a small store has been opened where the prime necessaries of life can be obtained. A fuller stock is kept at other stores from five to ten miles distant. Good building lumber retails at the mills, three, five and eight miles distant, at $8 to $12 per 1,000 feet, and dressed flooring at about $15.

In looking at the map issued by the State, you must bear in mind that there are thousands of beautiful little clear water lakes of from 10 to 400 acres, and a few of 400 to 3,000 not shown on it. Our locality has a goodly number of them. The proper way to reach our locality is the usual one "via Jacksonville, Fla." The St. Johns river steamers will carry you from Jacksonville to Astor, and then the railroad takes you to Fort Mason. About $3 will hire a private conveyance at Fort Mason, or near by at Lake Eustis post office, for this point—and then you will see one of the choice spots of Florida. We have no black gnats or sand flies, and a bar at night protects us from the mosquitoes during the few weeks in summer when we are subject to them. They never annoy us in daytime, as they do in some parts of New Jersey. There are less snakes on our uplands than in any unsettled part of the United States.

Drinking water is from our clear water lakes, springs and wells; it is soft and healthful. Boarding might be obtained among some of the families here for $12.00 per month, or $4.00 to $5.00 per week; and at the Bourland House, on Lake Ola, at reasonable rates. A post office has recently been established at Tangerine, and we have semi-weekly mail service. "The Tangerine Development Society" is an association composed chiefly of persons who have taken homesteads in this region, and who desire to see the merits of this locality more thoroughly made known to intending settlers. The officers of the society are D. W. Adams (P. M. National Grange), President; J. C. Russ, Vice President; R. J. Wright, Treasurer; J. H. Foster, Corresponding Secretary. For further particulars address the secretary.

LONGWOOD.

Longwood—Nine and a half miles from Sanford, on the South Florida Railroad, offers inducements to immigrants, and those wishing beautiful and healthful homes many advantages. It has been but a few years that the tide of seekers after Florida homes turned their steps this way, and now may be seen surrounding the beautiful clear water lakes, many large flourishing orange groves and comfortable homes.

Being on the line of the South Florida Railroad, we have rapid and convenient transportation for fruits, vegetables, lumber, &c. We have a railroad depot, a post office, two stores furnishing general supplies, a public school during the regular term, two churches, Episcopal and Baptist, a good boarding house, to which additions are to be made in time for winter travel, a first class carriage and wagon factory, conducted by a practical machinist, also two saw mills, turning out rough and planed lumber.

Our lands are high, rolling, and healthy; well adapted to the growing of all kinds of semi-tropical fruits and vegetables. Lying between the South Florida Railroad and the Wekiva river are the high, rolling pine hills of Orange County; a ridge running from the Sylvan Lake settlement on toward the southern portion of the county, healthy and salubrious, and rapidly being settled up by enterprising people from the North and West. Land in large and small lots can be had at reasonable rates by actual settlers. Letters of inquiry will be duly answered by addressing either of the following gentlemen:

F. H. RAND, S. F. GOVE,
E. L. CLARK, D. L. MILLER,
Longwood, Orange Co., Fla.

ALTOONA.

ALTOONA, in Orange County, is situated on the line of the St. Johns and Lake Eustis Railroad, seventeen miles from Astor on the St. Johns river, and eight miles from Fort Mason on Lake Eustis, and about the center of the most elevated section through which the said road leads. Altoona is one mile south of Lake Dorr, which is four miles long by three in width; this is one of the most beautiful bodies of water in western Orange.

The situation of Altoona is strikingly attractive, being on and among heavy rolling land, and surrounded by a number of beautiful clear water lakes. The soil is a sandy loam with a clay subsoil varying from one to four feet from the surface. This belt of country extends from the north of Lake Dorr to beyond Umatilla north and south for a distance of seven miles on the railroad, and from Acron to Niggertown Creek east and west, for a distance of twelve miles. This belt of country leading to the northeast and east of Lake Dorr is a high, rolling pine country, embracing about six thousand unoccupied acres of land taken by the State to supply the deficiency of the sixteenth section of the school land, which is an evidence of its superior quality.

There are here more young and flourishing groves, with many bearing, than in any other portion of our common county, with a well organized society, having the advantage of good church and school facilities. This land can be entered under the homestead law of the State, Acts of 1881, which is as follows: By actual residence and cultivation for three years, and paying for the land at three annual installments, one-third each year, the land being sold to settlers at one dollar and a quarter per acre. This land can be bought in lots from forty to one hundred and sixty acres.

The country directly east of Altoona, commencing two miles from same, is a flat country, interspersed with rising knolls of elevated land, many of which contain beautiful orange groves, and famous among them in this locality may be mentioned those of Dykes, Kirkland, Shultz and Crow. This section is also specially adapted to raising of stock, where they can be kept at no cost and little attention.

The county directly north of Altoona, and along the line of railroad to the head of Lake Dorr, is level but sufficiently elevated to insure perfect drainage and affording some of the best lands in this section.

The county west and north-west of Altoona is a high, hilly and rolling land, interspersed with many small, but beautiful clear water lakes, making it a very desirable locality for those in search of health. This section is being settled up rapidly by persons from the north and west.

To the south and south-west of Altoona towards Fort Mason for a few miles, the country is a beautiful specimen of rolling pine land, and in this section may be found some of the most beautiful lakes, surrounded by live oak, and affording some of the best land for general agriculture and semi-tropical fruits.

Citizens of this section are representatives of all parts of the Union, intelligent and hospitable. The healthfulness of this section is unsurpassed by any spot in the United States, water pure, sweet and as good as could be desired, though not as cold naturally as the water further north, owing to the evenness of the temperature of the climate. We have convenient schools for each neighborhood, under the management of efficient teachers, the church facilities are not as desired, but movements are being inaugurated to supply the deficiency. Our section is well supplied with steam saw mills, situated immediately on the railroad, which makes them accessible to each and every neighborhood.

ZELLWOOD.

ZELLWOOD—Latitude 28° 30' north, is situated in the beautiful lake region of West Orange County, twenty-five miles from the St. Johns river, four miles and a half from the Ocklawaha river, and ten miles southeast of Lake Eustis, the present terminus of the St. Johns and Lake Eustis Railroad. The topography of the country immediately surrounding is high, rolling and picturesque; the usual monotony of the piny woods being broken by hills, valleys and numerous clear water lakes. The soil, underlaid with clay and free from undergrowth, is of fair quality and well adapted to the production of all kinds of semi-tropical fruits. The location is perfectly healthy, and affords many desirable sites for groves and residences.

We have a school, post office, and a rapidly increasing population, largely composed of people of refinement and intelligence. The enlargement of the channels connecting Lakes Dora, Beauclair and Apopka, which is now rapidly progressing, will give us water transportation within three miles; while the railroad proposed from Lake Eustis to Orlando via Zellwood and Apopka will give us a speedy and convenient outlet in the near future. Visitors can best reach Zellwood by rail from Astor on the St. Johns to Lake Eustis, where conveyance can be obtained. Correspondence solicited.

GEORGE C. WELBY,
Zellwood, Orange Co., Fla.

Orange County Reporter

ORLANDO, ORANGE CO., FLA.

Mahlon Gore, - - - - - Publisher.

TERMS OF SUBSCRIPTION.—$2 per year; $1 for six months; 50 cents for three months. Single copies by mail for two 3-cent stamps.

The REPORTER is devoted to the development of South Florida, and aims in all things to keep fully abreast of the times in information concerning this favored region. No other portion of the United States is at present attracting so much attention as this; no other portion offers so many attractions to home-seekers, or capitalists desiring to invest money. It has the finest climate in the world, both in winter and summer; and statistics prove it to be the most healthy portion of the habitable globe. It is the home of the semi-tropical fruits, and only requires development to make it all that can be desired for an earthly home.

THE
SANFORD GRANT,
ORANGE COUNTY,
FLORIDA.

Land on this Grant, the property of The Florida Land and Coloñization Company, Limited, is now offered for sale in lots to suit purchasers, at prices varying, according to location and quality, on easy terms and long time to *actual settlers*. This property is, for the most part, an old Spanish Grant, confirmed under treaty with Spain by the United States Supreme Court, and, consequently, has a perfect title. It embraces nearly 25 square miles; and is situated in Orange County, on the south side of Lake Monroe, at the head of navigation for large steamers on the St. Johns River, in latitude 28 deg. 50 min. North. The population of the county has been increased ten-fold in as many years.

This section is now, without doubt, recognized as among the best in the State for the cultivation of the Orange and Lemon, by its exemption from injurious frosts, accessibility to market, and facilities for transportation. It is a notable fact that during the most severe cold the thermometer ranged 10° higher at Sanford, than 100 miles further north on the St. Johns River. The lowest range on the 30th of Dec., 1880, (the great freeze) was, at 6 a. m., 28°, and at 8 a. m., 32°, while it was 22° at Enterprise, across Lake Monroe north, and at Jacksonville and St. Augustine, 18°. At the State Fair at Jacksonville, on the 25th of January, the first prizes were given to pineapples, lemons, and to the lime, lemon and citron blossoms; cabbage, cauliflowers, turnips, &c., of Sanford—in all, nine premiums for the Grant—a unique exhibit after the frost which did so much damage in the Jacksonville region.

Its situation on the south side of Lake Monroe is very favorable for the growth of fruits, vegetables, &c., as it gives almost complete protection from frosts.

Sanford is a rapidly increasing incorporated town, with Shops, Churches, Reading Room, School, Money Order P. O., Daily Mails, a newspaper (South Florida Journal), Express and Telegraph, and the usual facilities of a growing town, and is the terminus of the South Florida Railroad, now completed to the head of the Kissimmee River, on the way to the Gulf of Mexico; it is the terminus for several lines of steamboats plying on the St. Johns River—one of them daily to and from Jacksonville; there is a steam saw, box and planing mill in the town, which supplies lumber at low prices; steam sash, door and blind factory; also car shops, in which all the cars of the S. F. R. R., with two or three exceptions, have been made.

There are fine openings for business of various kinds, especially factories for making paper, brushes, &c., of the palmetto; for utilizing the Spanish moss for mattresses and upholstery; for preserving and making jelly, marmalade, &c., of the guava, sweet and sour oranges, and other fruits. There is also an opening for a tannery—as annually large numbers of hides are sent North from this and adjoining counties for sale, while the very best tannin known can be had in unlimited quantities near at hand—and, in fact, this town offers good openings for all industries needed to supply a back country of over 15,000 population having its outlet here.

The Sanford House, one of the best hotels in the State, and numerous boarding houses, meet the requirements of travelers of every condition. Boating and fishing on Lake Monroe furnishes an unfailing resource for the pleasure seeker. A large, warm Sulphur Spring, within a mile of the hotel, possesses invigorating properties, while a cold spring of sulphur water supplies this healthful beverage to the invalid. An ever-flowing Artesian Well in the hotel grounds supplies the guests with an abundance of Sulphur Water, the beneficial effects of which are marked.

Lots in the town of Sanford for sale. Special rates and terms for the trades and those bringing new business enterprises.

APPLY TO

JAMES E. INGRAHAM, Agent,

Sanford, Orange County, Fla.

South Florida Railroad.

This road is now completed as far as

KISSIMMEE,

On Lake Tohoptalaga, the head waters of the Kissimmee River, and offers every facility to the Settler and to the

ORANGE, FRUIT OR VEGETABLE GROWER.

Its rates of freight and passage are lower than many of the old established roads. Its equipments are first-class, and its road bed one of the best in the State.

This Company has for sale, on reasonable terms.

LARGE BODIES OF LAND,

Well adapted to the culture of the fruits of the citrus family, as well as vegetables, cane, cotton, etc.

THE KISSIMMEE HOTEL,

AT LAKE TOHOPTALAGA,

The property of the Company, will be open for guests about the 15th of January, 1882.

All correspondence regarding lands should be addressed to the Chief Engineer,

E. R. TRAFFORD,

Sanford, Florida.

This town is the county seat of Orange County. It is located near the geographical center of the county. The region about it is a high pine country, dotted here and there with clear, fresh water lakes. The pine timber of this region is being rapidly cleared away and the ground set out in fruit trees. When all of the native timber shall give place to orange groves, and the rolling lands be embellished with tasteful residences, surrounded with groves and well kept grounds, this will be one of the most attractive portions of the State. The town is small, and although one of the oldest in the county, yet is essentially new. It is only during the past year that the town has sprung into life and vigorous growth. Prior to that time the land upon which the place is located was in litigation, and valid titles to the property could not be given. For this reason people who wished to locate in this vicinity were compelled to settle outside, and the country adjacent was developed far in advance of the town itself. With the settlement of the litigation came a demand for lots; new buildings were erected, new life seemed to pervade the community, and the place has taken on a vigorous and sturdy growth like that witnessed in many Western towns.

The completion of the South Florida Railroad from Sanford to Orlando, which was almost simultaneous with the ending of the litigation, contributed in no small degree to the boom which the town and surrounding country has enjoyed. This road has opened up transportation for fruit and vegetable products, and at the same time brought an army of enterprising home-seekers from other States, who, recognizing the fine attractions offered, have cast their lots here, and are contributing largely to the general development.

One of the most gratifying features of this new growth is the fact that these later accessions are from the most intelligent and substantial element of the several sections they represent. Society, which is now forming, promises to be of an exceptionally high order—socially, intellectually and morally.

The town is regularly incorporated and has a municipal organization. It has public schools during the public term with a total list of nearly one hundred and fifty pupils. Private schools are maintained during the months when there are no public schools, so that children can have the advantage of eight or nine months of school in the year. There are no church edifices as yet, although there are several regular church organizations. Baptists, Methodists, Episcopalians and Presbyterians are taking preliminary steps toward putting up church buildings, and the Catholics have secured ground for church and school in the near future. There are fifteen stores and business houses, covering the general range of merchandise, wagon and carriage factory, saddlery, livery stable, blacksmith and gunshops. A furniture factory and tannery are in contemplation. There are good hotels and boarding houses, but not of sufficient capacity to meet the requirements of the winter season. The town needs a large hotel. Cottages are in demand and hard to obtain. A large number of neat and convenient cottages would find ready tenents at good rates. Money judiciously invested in such buildings would give the owner satisfactory returns.

A pressing need of this section—as of the greater portion of the county—is lumber. Several mills are located near here, but they have not been able to keep up with the demand, and this scarcity of lumber has retarded building operations greatly. There need be no fears that the demand will be less for some years, as the rapid growth of town and country bids fair to increase rather than deminish.

Under judicious municipal regulations the taxes are low, and have steadily decreased for the past three or four years. Florida taxes are never burdensome. The South Florida Railroad is being extended southward to the head of navigation on the Kissimmee. This will give us direct rail and water communication with the Kissimmee and Okechobee regions. Other lines are projected both eastward and westward, and a few years at farthest will place Orlando on direct lines of communication with the Atlantic and Gulf coasts, and but three or four hours ride from either. This will insure a daily supply of salt water fish and oysters in the season, and make a "trip to the coast" but a morning ride.

One thing which is contributing largely to the prosperity and developement of this portion of the county is the fact that lands are being subdivided into small holdings of five, ten and twenty acres, and each of these small tracts is being taken for settlement and cultivation. People are beginning to understand that for fruit and vegetable culture in Florida, ten acres is enough. This system of subdivision insures better improvements, high culture and greater population.

Two practical nuserymen, of abundant means and extensive experience have purchased land adjacent to the town, and are now engaged in the preliminary work of these enterprises. This will be of great value to this section, as it will enable people to get supplies of trees at home, and at the same time give them the benefit of the intelligent experiments of practical horticulturists.

A private school of high grade, would be welcomed, and liberally patronized from the start. The manufacture of furniture would be profitable. A merchant tailor and a shoemaker are needed, and if competent would do well. The manufacture of marmalade from oranges and guavas is a field which is inviting attention, and orange wine from the cracked fruit, could be made in large quantities and at a handsome profit.

To those who are seeking homes in Florida. Orlando extends a cordial invitation to come and see for themselves. Come and see us. If you are pleased and wish to locate here, a hearty welcome awaits you.

Valuable Town Lots

FOR SALE

IN THE

BUSINESS CENTER

OF THE THRIVING

TOWN OF ORLANDO.

ALSO THE MOST VALUABLE

LOTS FOR DWELLINGS

AND

HOTEL PURPOSES.

FOR CASH OR ON TIME.

These lots have recently been

PLACED UPON THE MARKET

And are

FINDING A RAPID SALE FOR BOTH BUSINESS PURPOSES AND RESIDENCES.

The recent growth of the town is upon the tract of land of which these lots are a portion.

They are Rapidly Increasing in Value,

on account of the growing demand for them. For information as to prices and terms, apply to J. N. HARRINGTON, Orlando, Fla.

B. R. REID.

HON. D. F. HAMMOND.   E. M. HAMMOND.   JNO. C. JONES.

HAMMONDS & JONES,

Attorneys at Law

—AND—

Solicitors in Equity.

Will practice in all the courts,

STATE AND FEDERAL.

ALSO

REAL ESTATE AGENTS

Representing Finest Orange Groves and Un-cleared Lands in the State.

OFFICES AT

ORLANDO, ORANGE COUNTY, FLA.

New York Saddlery,

ORLANDO, FLA.

ESTABLISHED JANUARY, 1881.

BEST SHOP IN THE STATE.

A FULL LINE OF

HARNESS, SADDLES, BRIDLES, COLLARS, WHIPS, OIL, CHOICE BITS, AND SADDLERY HARDWARE.

BUGGY TRIMMINGS.

WAGON CUSHIONS OF LEATHER OR DUCK MADE TO ORDER.

ALL WORK DONE BY HAND.

No Factory Work or Shoddy Stock.

ORDERS BY MAIL PROMPTLY ATTENDED TO.

Call on or address,

NEW YORK SADDLERY, ORLANDO, FLORIDA.

MOST RELIABLE LINE EVER OFFERED IN SOUTH FLORIDA & AT NEW YORK PRICES.

Florida Curiosities, Optical Goods, &c.

A. G. SCHWERTER,

DEALER IN AND MANUFACTURER OF

WATCHES

CLOCKS, JEWELRY,

AND PLATED WARE,

ORLANDO, ORANGE COUNTY, FLA.

STATIONERY, &c.

REPAIRING of every description executed carefully and with dispatch. Orders by mail receive prompt attention. All goods warranted as represented, by written guarantee. To PERSONS AT A DISTANCE.—Any article you may desire in my line, send amount by Post Office Money Order or Registered Letter, and I will forward the same on trial. If not suitable I will refund amount on return of goods. Prices sent on application.

Alligator Teeth, Sea Bean & Shell Jewelry.

The geographical location of Orange County, lying as it does between the St. Johns River on the east, and the Ocklawaha River and Lakes Eustis, Dora, Beauclair and Apopka on the west, gives it not only unexcelled water transportation facilities, but protection from excessive cold, and renders it the natural home of the semi-tropical fruits, particularly so of the citrus family. The county embraces every variety of land: lands suited to fruit growing, vegetable gardening, sugar cane and general farming.

The healthfulness of the county is unexcelled by any country in the world.

Sanford, the head of navigation for large steamers on the St. Johns River, is in the center of the largest orange producing portion of the county. Within a radius of eight miles are 292 orange groves, in which are growing 165,625 orange and lemon trees, which will yield this year 2,568,000 oranges. Less than five per cent of the trees are yet bearing, the larger portion being young trees which will bear in from one to five years from this date.

Two railroads are completed and running in the county. The South Florida Railroad, which is completed from Sanford to the headwaters of the Kissimmee River, a distance of forty miles, runs through a thickly settled portion of the county.

Four other railroads will run into Sanford: the Sanford & Indian River Railroad; Sanford, Lake Eustis & Ocala Railroad; Tavares & Lake Monroe Railroad, and Palatka & Sanford Railroad, thereby making Sanford and its immediate vicinity the center of the railroad system of South and East Florida.

Improved and unimproved lands can be purchased at reasonable prices, adjoining and near to railroad and river transportation.

For particulars, &c., address,

J. J. HARRIS & CO.,
Sanford, Fla.

SAVANNAH, FLORIDA & WESTERN

RAILWAY,

OPERATING

THE WAYCROSS SHORT LINE

—AND—

Florida Dispatch Line.

THE SHORTEST, QUICKEST AND BEST ROUTE,

For Freight and Passenger Service,

TO ALL POINTS

NORTH, EAST AND WEST

From Orange County and all Points in Florida.

DOUBLE DAILY TRAIN SERVICE

To and From Jackonville and Callahan Junction.

SPECIAL ATTENTION PAID TO SHIPPING ORANGES,

Movement in Ventilated Cars on Fast Trains to destination by rail, without breaking bulk. Transfers by this line are at steamships' docks, without re-handling in transit. This is the only line that can give through dispatch of orange shipments to destination in the same cars that are loaded at the groves.

Rates Always as Low as the Lowest.

For full information on all matters pertaining to transportation by the New Line, address,

JAS. L. TAYLOR,
Gen. Freight and Passenger Agent.

H. S. HAINES,
General Manager.

D. H. ELLIOT, General Agent Florida Dispatch Line.